CHILDREN'S DEPT.

DISCARDED BY
MEMPHIS PUBLIC LIBRARY

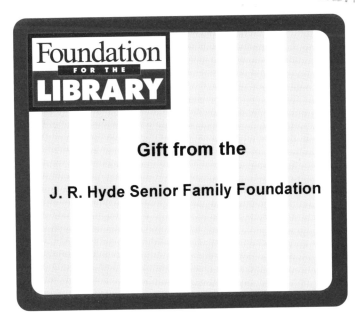

Gift from the

J. R. Hyde Senior Family Foundation

Let's Add
Bills

Kelly Doudna

Consulting Editor Monica Marx, M.A./Reading Specialist

Published by SandCastle™, an imprint of ABDO Publishing Company, 4940 Viking Drive, Edina, Minnesota 55435.

Copyright © 2003 by Abdo Consulting Group, Inc. International copyrights reserved in all countries. No part of this book may be reproduced in any form without written permission from the publisher. SandCastle™ is a trademark and logo of ABDO Publishing Company. Printed in the United States.

Credits
Edited by: Pam Price
Curriculum Coordinator: Nancy Tuminelly
Cover and Interior Design and Production: Mighty Media
Photo Credits: Eyewire Images, Hemera, PhotoDisc

Library of Congress Cataloging-in-Publication Data

Doudna, Kelly, 1963-
    Let's add bills / Kelly Doudna.
        p. cm. -- (Dollars & cents)
    Includes index.
    Summary: Shows how to use addition to find out how many dollars one needs to pay for various items and looks at different denominations of dollar bills, from one to one hundred.
    ISBN 1-57765-898-1
    1. Money--Juvenile literature. 2. Dollar, American--Juvenile literature. 3. Addition--Juvenile literature. [1. Money. 2. Addition.] I. Title. II. Series.

HG221.5 .D6519 2002
640'.42--dc21

2002071705

SandCastle™ books are created by a professional team of educators, reading specialists, and content developers around five essential components that include phonemic awareness, phonics, vocabulary, text comprehension, and fluency. All books are written, reviewed, and leveled for guided reading, early intervention reading, and Accelerated Reader® programs and designed for use in shared, guided, and independent reading and writing activities to support a balanced approach to literacy instruction.

Let Us Know

After reading the book, SandCastle would like you to tell us your stories about reading. What is your favorite page? Was there something hard that you needed help with? Share the ups and downs of learning to read. We want to hear from you! To get posted on the ABDO Publishing Company Web site, send us email at:

sandcastle@abdopub.com

SandCastle Level: Transitional

Bills are money.

one dollar
$1.00

five dollars
$5.00

ten dollars
$10.00

twenty dollars
$20.00

fifty dollars
$50.00

one hundred dollar
$100.00

We use bills to pay for things.

Let's see what we can buy.

The pizza costs $7.00.
$7.00 = 7 one-dollar bills

Jay has 5 one-dollar bills.

Jen has 2 one-dollar bills.

Do they have enough to buy the pizza?

Let's add.
5 + 2 = 7

The paintbrush costs $2.00.
$2.00 = 2 one-dollar bills

The paint costs $5.00.
$5.00 = 5 one-dollar bills

How many one-dollar bills does Lin need altogether?

Let's add.
2 + 5 = 7

The bike helmet costs $30.00.
$30.00 = 6 five-dollar bills

Sue has 2 five-dollar bills.

Tom has 4 five-dollar bills.

Do they have enough
to buy the bike helmet?

Let's add.
2 + 4 = 6

The soccer shoes cost $40.00.
$40.00 = 4 ten-dollar bills

The soccer ball costs $20.00.
$20.00 = 2 ten-dollar bills

How many ten-dollar bills does John need altogether?

Let's add.
4 + 2 = 6

The telescope costs $100.00.
$100.00 = 5 twenty-dollar bills

Brenda has 2 twenty-dollar bills.

Sam has 3 twenty-dollar bills.

Do they have enough to buy the telescope?

Let's add.
2 + 3 = 5

The bike costs $250.00.
$250.00 = 5 fifty-dollar bills

Lisa has 3 fifty-dollar bills.

Ken has 2 fifty-dollar bills.

Do they have enough to buy the bike?

Let's add.
3 + 2 = 5

The computer costs $400.00.
$400.00 = 4 hundred-dollar bills

Yuki has 2 hundred-dollar bills.

Dan has 2 hundred-dollar bills.

Do they have enough to buy the computer?

Let's add.
2 + 2 = 4

What are these bills called?

How much are they worth?

one dollar = $1.00
five dollars = $5.00
ten dollars = $10.00
twenty dollars = $20.00
fifty dollars = $50.00
one hundred dollars = $100.00

Index

bike, p. 17
bike helmet, p. 11
bills, pp. 3, 5, 21
computer, p. 19
dollar, p. 21
dollars, p. 21
fifty-dollar bills,
   p. 17
five-dollar bills,
   p. 11
hundred-dollar
   bills, p. 19
money, p. 3

one-dollar bills,
   pp. 7, 9
paintbrush, p. 9
paint, p. 9
pizza, p. 7
soccer ball, p. 13
soccer shoes, p. 13
telescope, p. 15
ten-dollar bills,
   p. 13
twenty-dollar bills,
   p. 15

Glossary

computer a machine that can process and store information

helmet a hard hat that protects your head

paint liquid color that you use to create pictures

paintbrush a brush used to spread paint

telescope an instrument used to look at things that are far away

About SandCastle™

A professional team of educators, reading specialists, and content developers created the SandCastle™ series to support young readers as they develop reading skills and strategies and increase their general knowledge. The SandCastle™ series has four levels that correspond to early literacy development in young children. The levels are provided to help teachers and parents select the appropriate books for young readers.

Emerging Readers
(no flags)

Beginning Readers
(1 flag)

Transitional Readers
(2 flags)

Fluent Readers
(3 flags)

These levels are meant only as a guide. All levels are subject to change.

To see a complete list of SandCastle™ books and other nonfiction titles from ABDO Publishing Company, visit **www.abdopub.com** or contact us at:
4940 Viking Drive, Edina, Minnesota 55435 • 1-800-800-1312 • fax: 1-952-831-1632